SpringerBriefs in Systems Biology

For further volumes:
http://www.springer.com/series/10426

M. V. K. Karthik · Pratyoosh Shukla

Computational Strategies Towards Improved Protein Function Prophecy of Xylanases from *Thermomyces lanuginosus*

 Springer

M. V. K. Karthik
Department of Biotechnology
Birla Institute of Technology
  (Deemed University)
Mesra Ranchi-835215
India

Pratyoosh Shukla
Department of Biotechnology
Birla Institute of Technology
  (Deemed University)
Mesra Ranchi-835215
India

and

Enzyme Technology and Protein
  Bioinformatics Laboratory
Department of Microbiology
Maharshi Dayanand University
Rohtak-124001 Haryana
India

ISSN 2193-4746   ISSN 2193-4754   (electronic)
ISBN 978-1-4614-4722-1   ISBN 978-1-4614-4723-8   (eBook)
DOI 10.1007/978-1-4614-4723-8
Springer New York Heidelberg Dordrecht London

Library of Congress Control Number: 2012942259

Printed on acid-free paper

Springer is part of Springer Science+Business Media (www.springer.com)

Contents

Abstract

Thermophilic xylanases are vital for numerous processes activated at high temperatures including their application in pulp and paper industries for biobleaching. Here, we proclaim that mutants of a family 11 xylanases from Thermomyces lanuginosus with a single point mutations viz.,Y77W, Y77R, and Y77F are constructed from Tyr77 residues in active sites of wild strains. The accuracy of such mutant models was verified by Ramachandran plot (84.7 % amino acids in most favoured region; 15.3 % in additionally allowed regions) and ERRAT studies (accuracy of 89.13 %). Further, endo-1,4-beta-xylanase(1YNA) and 19 model mutants were docked with xylobiose and beta-D-xylopyranose as substrates revealing that Y77W, Y77R, and Y77F are having mutant residues in their binding sites showing the docking scores of −272.50 (Y77W), −272.65(Y77R), −265.81(Y77F) and −171.47(Y77W),−174.71(Y77R), −170.63(Y77F) for xylobiose and beta-D-xylopyranose, respectively indicating their higher stability. Further from automatic protein function prediction system (3D2GO), it is proven that Y77W is more stable with highest molecular function (0.58) than the other two mutants i.e., Y77R (0.52) and Y77F (0.51) as well as to wild type (0.46) also. Interestingly, it was also noted that catalytic activity was maximum in mutant Y77F (0.95). These kinds of strategies can be used to improve the efficiency of optimization processes of enzyme stability and rational design of thermophilic xylanases.

Chapter 1
Introduction

1.1 Introduction

Computational protein design assists methodical, high-throughput protein and ligand mutagenesis and has been the focus of several researchers in recent years, resulting in significant improvements in methodology and application. It is evident that computational protein design, has been used to transform specificity (Bolon et al. 2005; Ashworth et al. 2006; Green et al. 2006; Lopes et al. 2010), improve protein-ligand binding (Klepeis et al. 2003; Shifman and Mayo 2003; Lippow et al. 2007; Altman et al. 2008; Reynolds et al. 2008; Haidar et al. 2009; Bellows et al. 2010), amplify stability (Korkegian et al. 2005; Shah et al. 2007), stabilize novel or substitute protein folds (Kuhlman et al. 2003; Ambroggio and Kuhlman 2006), design new proteins (Calhoun et al. 2003; Calhoun et al. 2008), and deciphering enzyme active sites (Tynan-Conolly and Nielsen 2007; Röthlisberger et al. 2008). These techniques also play vital role in optimizing ligand entrance and escape redundant pathways (Chaloupkova et al. 2003), re-modeling of protein-protein interfaces (Joachimiak et al. 2006; Grigoryan et al. 2009), and rewiring biological network (derSloot et al. 2009).

In this work, we persistently describe the computational design of family 11 xylanase (endo-1, 4-β-xylanase) from Thermomyces lanuginosus showing enhanced functions for instance hydrolase activity, proteolysis, and binding activity. Endo-1, 4-β-xylanase generally called 1, 4-β-D-xylanxylanohydrolase (EC 3.2.1.8) cleaves the glycosidic bonds in the xylan backbone, bringing about a reduction in the degree of polymerization of the substrate (Al Balaa et al. 2009). It was perceptible that a good understanding of the molecular basis for enhanced effect of these mutant forms may be encouraging for rational design and engineering of xylanases. In this study, we report a single-point mutation (You et al. 2009) in the active center of a family 11 xylanase (endo-1, 4-β-xylanase) resulting significant improvement of hydrolase

M. V. K. Karthik and P. Shukla, *Computational Strategies Towards Improved Protein Function Prophecy of Xylanases from* Thermomyces lanuginosus, SpringerBriefs in Systems Biology, DOI: 10.1007/978-1-4614-4723-8_1, © The Author(s) 2012

activity, proteolysis, and binding activity in mutants. Further exploration of the molecular basis for such improved effects will be crucial while designing enzyme optimization processes.

References

Al Balaa B, Brijs K, Gebruers K, Vandenhaute J, Wouters J, Housen I (2009) Xylanase XYL1p from scytalidium acidophilum: site-directed mutagenesis and acidophilic adaptation. Biores Technol 100:6465–6471

Altman MD, Nalivaika EA, Prabu-Jeyabalan M, Schiffer CA, Tidor B (2008) Computational design and experimental study of tighter binding peptides to an inactivated mutant of HIV-1 protease. Proteins 70:678–694

Ambroggio XI, Kuhlman B (2006) Computational design of a single aminoacid sequence that can switch between two distinct protein folds. J Am Chem Soc 128:1154–1161

Ashworth J, Havranek JJ, Duarte CM, Sussman D, Monnat RM Jr, Stoddard BL, Baker DL (2006) Computational redesign of endonuclease DNA binding and cleavage specificity. Nature 441:655–659

Bellows ML, Fung HK, Taylor MS, Floudas CA, devictoria AL, Morikis (2010) New compstatin variants through two de novo protein design frameworks. Biophys J 98:2337–2346

Bolon DN, Grant RA, Baker T, Sauer RT (2005) Specificity versus stability in computational protein design. Proc Natl Acad Sci USA 102:12724–12729

Calhoun JR, Kono H, Lahr S, Wang W, deGrado WF, Saven JG (2003) Computational design and characterization of a monomeric helical dinuclearmetalloprotein. J Mol Biol 334:1101–1115

Calhoun JR, Liu W, Spiegel K, Peraro MD, Klein ML, Valentine KG, Wand JA, deGrado WF (2008) Solution NMR structure of a designed metalloprotein and complementary molecular dynamics refinement. Structure 16:210–215

Chaloupkova R, Sykorova J, Prokop Z, Jesenska A, Monincova M, Pavlova M, Tsuda M, Nagata Y, Dambrovsky J (2003) Modification of activity and specificity of haloalkane dehalogenase from sphingomonas paucimobilis UT26 by engineering of its entrance channel. J Biol Chem 278:52622–52628

derSloot AMV, Kiel C, Serrano L, Stricher F (2009) Protein design in biological networks: from manipulating the input to modifying the output. Prot Eng Des Sel 22:537–542

Green DF, Dennis AT, Fam PS, Tidor B, Jasanoff A (2006) Rational design of new binding specificity by simultaneous mutagenesis of calmodulin and a target peptide. Biochemistry 45:12547–12559

Grigoryan G, Reinke AW, Keating AE (2009) Design of protein interaction specificity gives selective bZIP-binding peptides. Nature 458:859–864

Haidar JN, Pierce B, Yu Y, Tong WW, Li M, Weng ZP (2009) Structure-based design of a T-cell receptor leads to nearly 100-fold improvement in binding affinity for pepMHC. Proteins 74:948–960

Joachimiak LA, Kortemme T, Stoddard BL, Baker D (2006) Computational design of a new hydrogen bond network and at least a 300-fold specificity switch at a protein-protein interface. J Mol Biol 361:195–208

Klepeis JL, Floudas CA, Morikis D, Tsokos CG, Argyropoulos E, Spruce L, Lambris JD (2003) Integrated computational and experimental approach for lead optimization and design of compstatin variants with improved activity. J Am Chem Soc 125:8422–8423

Korkegian A, Black ME, Baker D, Stoddard BL (2005) Computational thermo stabilization of an enzyme. Science 308:857–860

Kuhlman B, Dantas G, Ireton GC, Varani G, Stoddard BL, Baker DL (2003) Design of a novel globular protein fold with atomic-level accuracy. Science 302:1364–1368

Lippow SM, Wittrup KD, Tidor B (2007) Computational design of antibody-affinity improvement beyond in vivo maturation. Nat Biotechnol 25:1171–1176

Lopes A, Schmidt, Busch M, Simonson T (2010) Computational design of protein-ligand binding: modifying the specificity of asparaginyl-tRNAsynthetase. J Comp Chem 31:1273–1286

Reynolds KA, Hanes MS, Thomson JM, Antczak AJ, Berger JM, Bonomo RA, Kirsch JF, Handel TM (2008) Computational redesign of the SHV-1 beta-lactamase/beta-lactamase inhibitor protein interface. J Mol Biol 382:1265–1275

Röthlisberger D, Khersonsky O, Wollacott AM, Ziang L, Dechncie J, Betker J, Gallaher JL, Althoff EA, Zanghellini A, Dym O, Albeck S, Houk KN, Tawfik DS, Baker D (2008) Kemp elimination catalysts by computational enzyme design. Nature 453:190–195

Shah PS, Hom GK, Ross SA, Lassila JK, Crowhurst KA, Mayo SL (2007) Full-sequence computational design and solution structure of a thermostable protein varian. J Mol Bio 372:1–6

Shifman JM, Mayo SL (2003) Exploring the origins of binding specificity through the computational redesign of calmodulin. Proc Natl Acad Sci USA 100:13274–13279

Tynan-Conolly BM, Nielsen JE (2007) Redesigning protein pKa values. Prot Sci 16:239–249

You C, Yuan H, Huang Q, Lu H (2009) Substrate molecule enhances the thermostability of a mutant of a family 11 xylanase from Neocallimastix patriciarum. Afr J Biotechnol 9:1684–5315

Chapter 2
Background

2.1 Protein Structure

Proteins are an important class of biological macromolecules present in all biological organisms. Proteins consist of a sequence of 20 different amino acids, also referred to as residues. To be able to perform their biological function, proteins often fold into one, or more, specific spatial conformations, driven by a number of non-covalent interactions, such as hydrogen bonding, ionic interactions, Van der Waals' forces, and hydrophobic packing. In order to understand the functions of proteins at a molecular level, it is often necessary to determine their 3D structure. This is the topic of the scientific field of structural biology that employs techniques, such as X-ray crystallography, nuclear magnetic resonance (NMR) spectroscopy, and electron microscopy, to determine the structure of proteins.

The number of known protein structures deposited in the Protein Data Bank (PDB) has grown exponentially over the last 30 years. This trend can be expected to continue as structural genomics projects gain momentum and techniques allowing higher throughput structure determination are developed. Currently there are more than 40,000 crystallographic or NMR structures of proteins or nuclear acids available in PDB.

2.1.1 Classification of Protein Structure

Several methods of structural classification of proteins have been developed to introduce some order to the large amount of data present in the PDB. Such methods facilitate structural comparisons and provide a greater understanding of structure and function (Hadley and Jones 1999). The most widely used and comprehensive classification databases are structural classification of proteins (SCOP) (Murzin

M. V. K. Karthik and P. Shukla, *Computational Strategies Towards Improved Protein Function Prophecy of Xylanases from* Thermomyces lanuginosus, SpringerBriefs in Systems Biology, DOI: 10.1007/978-1-4614-4723-8_2, © The Author(s) 2012

et al. 1995), class architecture topology homologous (CATH) (Orengo et al. 1997). These classifications make use of different methods of defining and categorizing protein folds that lead to different views of the protein-fold space.

SCOP was among the earliest efforts to classify protein structures into folds. Protein domains with no obvious sequence homology to other domains are defined and classified manually. In many ways, this database has been considered the standard for protein structure classification.

CATH superfamily makes use of a combination of manual and automated procedures in defining and classifying protein domains. CATH relies on the consensus of three automated classification methods to break protein chains into domains.

2.1.2 Protein Surface

The protein surface is the outer or the topmost boundary of a protein. The topology of the surface of a protein is intimately related to its function; parts of the surface are directly involved in interactions with other molecules; the solvent-protein interface is almost certainly related to the structure of the native molecule; and the chemical reactivity of the various functional groups will depend on their relation to this interface (Lee and Richards 1971).

There are three types of protein surfaces: van der Waals surface, solvent excluded surface (Connolly surface), and solvent accessible surface. van der Waals surface corresponds to the envelope containing the atomic spheres of van der Waals radius. The shape of the van der Waals surface of a molecule may be misleading, especially for macromolecules, since it frequently contains small gaps, pockets, and clefts which are sometimes too small to be penetrated even by a solvent molecule like water. For all practical purposes, the van der Waals surface of these oddments cannot enter into contact with a solvent or a drug molecule and therefore is not truly an accessible surface. To smooth the roughness of the van der Waals surface, Lee and Richards (1971) introduced the concept of a contact surface and a solvent accessible surface. These surfaces are obtained by rolling a spherical probe of a diameter corresponding to the size of a solvent molecule (usually water) on the original van der Walls surface. As a result, the area where the probe touches the van der Waals surface is called the contact surface, the center of the spherical probe traces a surface called the solvent accessible surface and the patches over narrow gaps and clefts traced by the surface of the probe are called re-entrant surfaces. The Connolly surface is composed of contact surface and re-entrant surface.

2.2 Protein–Protein Interaction

Proteins do not live alone; they have to interact with other molecules like DNA and RNA as well as other proteins to perform their biological function. Protein interactions play a critical role in all stages of cellular development, metabolism, and biological pathways. For example, signals from the exterior of a cell are mediated to the inside of that cell by protein–protein interactions of the signaling molecules. This process, called signal transduction, plays a fundamental role in many biological processes and in many diseases (e.g. cancer). Proteins might interact for a long time to form part of a protein complex, a protein may be carrying another protein (for example, from cytoplasm to nucleus or vice versa in the case of the nuclear pore importins), or a protein may interact briefly with another protein just to modify it (for example, a protein kinase will add a phosphate to a target protein). This modification of proteins can itself change protein–protein interactions. The specific interactions between proteins are critical in most cellular processes. The normal function of a cell can be seriously damaged by the disruption of interactions or by the non-specific aggregation with other proteins.

In a word, protein–protein interactions are of great importance for virtually every process in a living cell. Information about these interactions improves our understanding of diseases and can provide the basis for new therapeutic approaches. There are many methods developed either from biological experimental or computational perspective, as will be discussed.

There are several biological experimental techniques developed in the last two decades to detect the interactions between proteins. One of the most important techniques is the yeast two-hybrid system. Two-hybrid screening is a molecular biology technique used to discover protein–protein interactions (Criekinge and Beyaert 1999; Young 1998) and protein-DNA interactions by testing for physical interactions (such as binding) between two proteins or a single protein and a DNA molecule, respectively. Phizicky and Fields (1995) and Shoemaker and Panchenko (2007) have given a good review of experimental methods to investigate protein–protein interactions.

Although high-throughput experimental methods produce a large amount of data about protein interactions, interactomes of many organisms are far from complete. The low interaction coverage along the experimental biases toward certain protein types and cellular localizations reported by most experimental techniques call for the development of computational methods to predict whether two proteins will interact (Shoemaker and Panchenko 2007). The problem of predicting protein–protein interactions computationally has been tackled by a number of groups in different fields of structural and functional genomics. From a structural perspective, protein–protein interface studies have perhaps been the most successful computational approaches. Structural analysis of the interfaces of known protein–protein interactions allows common structural interaction motifs to be identified. These motifs can be used to predict whether it is possible for two proteins of known structure to physically interact (Kim et al. 2004). Some computational methods are based on the colocalization of potentially interacting genes in the same gene clusters or protein chains (gene cluster (Overbeek et al.1999), gene neighborhood (Galperin and Koonin 2000; Huynen et al. 2000; Rogozin et al. 2002), and Rosetta stone methods (Marcotte and Marcotte 2002; Marcotte et al. 1999; Yanai et al. 2001), on co-evolution patterns in interacting proteins (sequence co-evolution methods (Bowers et al. 2004; Goh et al. 2000; Jothi et al. 2006; Kim et al. 2004), and on the co-expression of genes. Some methods find patterns of co-occurrences in interacting proteins, protein domains, and phenotypes phylogenetic profiles (Pellegrini et al. 1999; Snitkin et al. 2006), while others use the presence of sequence/structural motifs characteristic only for interacting proteins.

2.3 Protein Binding Site Prediction

As discussed above, proteins have to interact with other molecules like DNA, small molecules (ligand), or other proteins to perform their biological function. Knowledge about where the protein binds to other molecules gives us a better understanding of its biological function. Before discussing protein–protein inter-action site prediction, we will discuss the computational approaches to predict protein–ligand binding site.

2.3.1 Identification of Protein–Ligand Binding Site

Proteins not only interact with other proteins but also interact with some small molecules (called ligands here) like NAD (Nicotinamide adenine dinucleotide), AMP (Adenosine 5′-monophosphate), etc. Unlike the interaction between proteins, ligands tend to bind to the pockets (cavities) on protein surface. Identification and evaluation of these ligand binding sites are the initial steps for protein structural-based drug design. Characterizing these ligand binding pockets plays an important role in automated ligand docking.

In the last decade, a variety of computational methods has been developed for the location of possible ligand-binding sites of proteins. Most of these pocket detection methods use pure geometric criteria to find clefts on protein surface and do not require any knowledge of the ligands, such as POCKET (Levitt and Banaszak 1992), LIGSITE (Hendlich et al. 1997), SURFNET (Laskowski 1995), CAST (Liang et al. 1998), and PASS (Brady and Stouten 2000). Statistical and empirical studies have shown that the actual ligand binding sites correspond to the largest pocket on a protein surface (Laskowski 1995; Laskowski et al. 1996).

One of the first methods, POCKET (Levitt and Banaszak 1992), introduced the idea of protein-solvent-protein events as key concept for the identification. The protein is mapped onto a 3D grid. A grid point is part of the protein if it is within 3 A of an atom coordinate; otherwise it is solvent. Next, the x, y, and z-axes are scanned for pockets, which are characterized as a sequence of grid points, which start and end with the label protein and a period of solvent grid points in between. These sequences are called protein-solvent-protein events. Only grid points that exceed a threshold of protein-solvent-protein events are retained for the final pocket prediction. Since the definition of a pocket in POCKET is dependent on the angle of rotation of the protein relative to the axes, LIGSITE extends POCKET by scanning along the four cubic diagonals in addition to the x, y, and z directions.

In SURFNET (Laskowski 1995), the key idea is that a sphere, which separates two atoms and which does not contain any atoms, defines a pocket. First, a sphere is placed so that the two given atoms are on opposite sides on the sphere's surface. If the sphere contains any other atoms, it is reduced in size until no more atoms are contained. Only spheres, which are between a radius of 1–4 A are kept. The result of this procedure is a number of separate groups of interpenetrating spheres, called gap regions, both inside the protein and on its surface, which correspond to the protein's cavities and clefts (Laskowski et al. 1996).

CAST (Binkowski et al. 2003; Liang et al. 1998) computes a triangulation of the protein's surface atoms using alpha shapes. In the next step, triangles are grouped by letting small triangle flow toward neighboring larger triangles, which act as sinks. The pocket is then defined as collection of empty triangles. CAST was tested on 51 of 67 enzyme-ligand complexes used for SURFNET (Laskowski et al. 1996) and achieved a success rate of 74 %.

PASS (Brady and Stouten 2000) uses probe spheres to fill cavities layer-by-layer. First, an initial coating of the protein with probe spheres is calculated. Each probe has a burial count, which counts the number of atoms within 8 A distance. Only probes with count above a threshold are retained. This procedure is iterated until a layer produces no new buried probe spheres. Then each probe is assigned a probe weight, which is proportional to the number of probe spheres in the vicinity and the extent to which they are buried. Finally, a small number of active site points (ASP) are selected by identifying the central probes in regions that contain many spheres with high burial count. The final ASP are determined by cycling through the probes in descending order of probe weight, keeping only those above a threshold and farther than 8.0 A apart from each other. Finally, the retained ASP are ranked by probe weight.

The largest pocket identified by these geometric methods on the protein surface is usually the binding site of a ligand. Besides these purely geometric methods above, there are methods, which take additional information into account to re-rank predictions. Glaser et al. (2006) refined SURFNET's predictions by considering the degree of residue conservation in the pocket. Q-SITEFINDER (Laurie and Jackson 2005) used the interaction energy between the protein and a simple van der Waals probe to locate energetically favorable binding sites. Eyrisch and Helms (2007) proposed a novel pocket detection protocol and applied it on three selected proteins BCL-XL, IL-2, and MDM2, to identify transient pockets which could be inhibitor binding sites. The unbound structures were used as starting points for 10 ns long molecular dynamics simulations. Then trajectory snapshots were scanned for cavities on the protein surface using the program PASS (Brady and Stouten 2000). The inhibitors were placed into the transient cavities they detected using AutoDock (Morris et al. 1998) and the complexes were compared to known inhibitor bound complexes with less than 2 °A RMSD. The results showed that this protocol could be a viable tool to identify transient ligand binding pockets on protein surfaces (Eyrisch and Helms 2007).

As discussed above, there are many pocket identification methods developed to predict protein–ligand binding site. However, they cannot be compared directly since different methods use different test datasets, different representations of pocket sites, and different assessment methods.

2.3.2 Protein–Protein Binding Site Prediction

Protein–protein interactions play a critical role in protein function. Completion of many genomes is being followed rapidly by major efforts to identify experimentally interacting protein pairs in order to decipher the networks of interacting, coordinated-in-action proteins. Identification of protein–protein interaction sites and detection of specific residues that contribute to the specificity and strength of protein interactions is an important problem (Chothia and Janin 1975; Yan et al. 2004) with broad applications ranging from rational drug design to the analysis of metabolic and signal transduction networks. Computational efforts to identify protein–protein interaction sites, in particular, identify surface residues that are associated with protein–protein interaction, play an increasingly important role because the experimental determination of protein structures, protein–protein complexes lag behind the number of protein sequences. During the last decade, many efforts have been made to analyze the properties of protein–protein interaction in order to predict interaction sites (Bradford and Westhead 2005; Jones and Thornton 1997; Neuvirth et al. 2004; Yan et al. 2004; Zhou and Shan 2001).

2.3.2.1 Surface Patch-Based Prediction Methods

To analyze the surface properties, the surface is usually divided into several patches and these patches have different characteristics which can be a measure to distinguish the binding sites from the rest of the surface. In 1997, Jones and Thornton analyzed the surface patches using six parameters: solvation potential, residue interface propensity, hydrophobicity, planarity, protrusion, and solvation accessible surface area (ASA). The six parameters were then combined into a global score that gave the probability of a surface patch forming protein–protein interaction.

Fernandez-Recio et al. (2004) applied protein docking simulations and analysis of the interaction energy landscapes to identify protein–protein interaction sites. The ensembles of the solutions generated by the simulations were subsequently used to project the docking energy landscapes onto the protein surface. They found that highly populated low-energy regions consistently corresponded to actual binding sites. In their results, as much as 81 % of the predicted high propensity patch residues were located correctly in the native interface. An obvious shortcoming of this approach is its slow speed since it is very time-consuming to do docking simulations to a large data set. Moreover, the same authors proposed optimal docking area (ODA), a method of analyzing a protein surface in search of areas with favorable energy change when buried upon protein–protein association (Recio et al. 2005). This method identified continuous surface patches with optimal docking solvation energy based on atomic solvation parameters. This energy was calculated using an atomic ASA-based model according to the following equation:

$$E\,\text{desov} = -\Sigma\sigma\,i\,\text{ASA}i$$

where $\sigma\,i$ is the atomic solvation parameter (ASP) for atom type i (i.e. the contribution to the solvation energy per unit of ASA).

Liang et al. (2004) developed a new energy scoring function and applied it to the analysis of the surface patches. They found that the patch with the highest energy score overlapped with the observed interface and the residue with the highest energy score of a small promoter was very likely the key interaction residue. Neuvirth et al. (2004) developed a method called ProMate to distinguish interface regions based on 13 properties. They got a success rate of 65 % on a test data set of 57 proteins. Liang et al. (2006) present an empirical scoring function PPINUP, which is a linear combination of energy score, interface propensity, and residue conservation score for the prediction of protein–protein binding sites.

2.3.2.2 Machine Learning Prediction Approaches

The support vector machine (SVM) is a supervised learning algorithm, useful for recognizing subtle patterns in complex datasets. The algorithm performs discriminative classification, learning by example to predict the classifications of previously unseen data. The algorithm has been applied in domains as diverse as text categorization, image recognition, and hand-written digit recognition. Recently SVM has been used for the prediction of protein interaction interfaces (Bradford and Westhead 2005; Koike and Takagi 2004; Yan et al. 2004). Yan et al. (2004) generated a SVM classifier to determine whether or not a surface residue is located on the interface using information about the sequence neighbors of a target residue. In their leave-one-out experiment, the SVM classifier was trained using a set of surface residues from a combined set of 115 protein complexes, labeled as interface or non-interface. Their prediction results showed that SVM yields relatively high sensitivity (0.51) and specificity (0.41).

As mentioned above, no single parameter absolutely differentiates interface from other surface patches. In the method of PPI PRED, a SVM was trained to distinguish between interacting and non-interacting surface patches using six of the surface properties: surface shape, hydrophobicity, conservation, electrostatic potential, residue interface propensity and solvent ASA (Bradford and Westhead 2005). Then this SVM was used to predict interface surface patches of proteins not included in the training set. Using this method, PPI PRED was able to successfully predict the location of the binding sites on 76 % of the 180 protein data set using a leave-one-out validation procedure. PPI PRED was applicable to both obligate and transient binding sites.

Chen and Zhou (2005) developed a method called PPISP to predict protein–protein interaction sites from a neural network with sequence profiles of neighboring residues and solvent exposure as input. The network was trained on 615 pairs of non-homologous complex-forming proteins. Tested on a different set of 129 pairs of non-homologous complex-forming proteins, 70 % of the 11,004 predicted interface residues are actually located in the interfaces. These 7,732 correctly predicted residues account for 65 % of the 11,805 residues making up the 129 interfaces. The main strength of the network predictor lied in the fact that neighbor lists and solvent exposure are relatively insensitive to structural changes accompanying complex formation. Fariselli et al. did a similar job, but their neural networks were trained with a reduced representation of the interacting patches and the success rate slightly increased to 73 %. SPPIDER is a neural network prediction method that includes predicted relative solvent accessibility as input (Porollo and Meller 2007). Sen et al. (2004) developed a consensus methodology which combines four different methods: data mining using Support Vector Machines, threading through protein structures, prediction of conserved residues on the protein surface by analysis of phylogenetic trees, and the Conservatism of Conservatism method of Mirny and Shakhnovich (1999). Their prediction results on a dataset of 7 hydrolase-inhibitor complexes demonstrated that the combination of all four methods improved predictions over the individual methods. Recently, proposed to use conditional random fields (CRFs) to predict protein interaction sites by solving sequential labeling problem with features including the protein sequence profile and the residue accessible surface area. The authors compared this CRFs-based method with the other methods such as support vector machines and neural networks. The comparative experiments on 1,276 non-redundant chains of hetero complexes showed that CRFs-based method achieved the best performance on a complete data set.

2.4 Protein–Protein Docking

Practically every process in the living cell requires molecular recognition and formation of complexes, which may be stable or transient assemblies of two or more molecules with one molecule acting on the other, or promoting intra- and intercellular communication, or permanent oligomeric ensembles (Eisenstein and Katchalski-Katzir 2004). The rapid accumulation of data on protein–protein interactions, sequences, structures calls for the development of computational methods to help our understanding of live cells. One of the methods is involved in the prediction of the complex structure from its components. Typically docking methods are investigated which attempt to predict the complex structures given the structures of components. Over the past 30 years many docking approaches have been proposed, ranging from thermodynamic approaches to correlation approaches, from rigid body docking to flexible docking.

Docking algorithms operate on the atomic coordinates of two individual proteins usually considered as rigid bodies and generate a large number of candidate association models between them. These candidates are then ranked by using various scoring functions, used independently or in combination. The scoring functions generally include geometric and chemical complementarities measures, electrostatics, hydrogen-bonding interaction and van der Waals interaction, and some empirical potential functions. A number of algorithms and many different scoring functions have been developed in the last ten years, as recently reviewed by Eisenstein and Katchalski-Katzir (2004), Halperin et al. (2002), Vajda and Camacho (2004), and the field has become extremely active.

2.4.1 Rigid-Body Docking

In the rigid-body docking approaches, the proteins are considered as rigid and no flexibility is taken into account. Here, an overview is given of the different steps involved in rigid-body protein–protein docking:

1. Start with the experimented 3D structures of the two unbound component proteins. Assuming that the formed complex has limited conformational changes, the two component proteins are regarded as rigid-bodies.
2. A 6D rotational and translational degree search is performed over all possible associations since in most cases of unbound–unbound complexes there is no biological information about which parts of the proteins will interact. This search will sample the space of all possible associations and consequently there will be a lower limit on the difference in conformations between two docked predicted complexes that determines the solution of the search procedure.
3. A large number of different complexes are generated after the global search procedure. Then a function is developed to score the quality of these docked (predicted) complexes. At this stage, geometric or electrostatic complementarity is often used since it is very fast to compute. Ideally, the docking algorithms thereby identify several complexes which are close to the native complex based on these complexes having best score.
4. Ideally, the docking algorithms thereby identify several complexes which are close to the native complex based on these complexes having best score.
5. If the experimental complex structure is known, then the predicted complex structure is superimposed and the root mean square deviation (RMSD) between all $C\alpha$ atoms of the predicted and native structure is calculated to evaluate the quality of the docking method. The predicted one can be regarded as near-native complex structure if the RMSD is below 3 Angstrom.
6. Then a re-ranking of the resultant complexes can be undertaken possibly using computational more intensive calculations.

7. Finally, conformational flexibility can be introduced into the algorithm to refine
   the few predicted complexes structure when there are only a limited number of
   complexes to consider.

Starting from two unbound structures, a lot of docked complex structures are
generated by a structure generator. Then these structures are filtered using a scoring
function and only a few favorable structures are left for evaluation in more details.

2.4.2 Fast Fourier Transform

In the first step of many docking methods, an attempt is made to represent the
protein structures in an efficient way. In 1992, Katchalski-Katzir et al. first proposed fast Fourier transform method (FFT) which were further developed by
several authors (Ben-Zeev and Eisenstein 2003; Chen and Weng 2003; Eisenstein
and Katchalski-Katzir 2004, Gabb et al. 1997; Sternberg et al. 1998; Tovchigrechko et al. 2002). A good review about this method can be found in Eisenstein and Katchalski-Katzir (2004).

Thus, for molecule a surface grid points are given the value 1, those in the
interior are given the value p (usually-15), and grid points outside the molecule are
given a value of 0. For molecule b, grid points on the surface and in the interior of
the molecule are given the value q (usually 1).

These two grids can then be superimposed and the mobile grid (protein B) is
translated by shifts α, β, γ. The value of a $l,m,n \times b\ l - \alpha,m - \beta,n - \gamma$ gives the
extent of shape complementarity for grid point (l, m, n) of the grid of protein A.
A value of 1 for the product grid indicates that the cell of protein B is superimposed
on the surface of A which indicates favorable shape complementarity. A value of
-15 indicates a steric clash with the cell from B superimposed on the core of A. The
value of -15 is chosen to penalize but not totally prevent steric clashes. A value of
zero means that there is no overlap between the two proteins. Thus, the totally
shaped complementarity for the two superimposed grid c α,β,γ is calculated from

$$C_{\alpha,\beta,\gamma} = \sum_{l=1}^{N}\sum_{m=1}^{N}\sum_{n=1}^{N} a_{l,m,n}\, Xb_{l-\alpha,m-\beta,n-\gamma}$$

The value of c is a convolution and its calculation requires O (N 6) complexity.
To reduce the calculation time, we use the discrete FFT to speed up the process of
calculation. A discrete FFT is

$$DFT\left(A_{l,m,n}\right) = \sum_{l=1}^{N}\sum_{m=1}^{N}\sum_{n=1}^{N} \exp[-2\pi i(pl + qm + rn)/N]\, Xa_{l,m,n}$$

First, calculate the discrete FFT for grid A and grid B. Then, calculate the complex conjugate of DF T (A) which is denoted as DF T − 1 (A). Then by Fourier theory, the discrete Fourier transform of c, DF T (C) is given by

$$DF\ T\ (C)\ =\ DF\ T - 1(A)DF\ T\ (B)$$

Therefore, c = IF T [DF T (C)] where IF T is the inverse Fourier transform.

In summary, this discrete Fourier transform can reduce the complexity of $O(N\ 6)$ to $O(N\ 3\ logN\ 3)$ (N is the size of the grid) (Katchalski-Katzir et al. 1992). Moreover, the imaginary part in a complex can be used to store additional complementarity information. For example, electrostatics complementarity and hydrophobic complementarity (Berchmanski et al. 2002) are added to filter the predicted complex structures together with geometric shape complementarity.

2.4.3 Benchmark for Testing the Docking Algorithms

The docking problem can be divided into two classes depending on the input of component structures. If we separate the complex structure into two components and then try to dock them together, it is called bound–bound docking. This is quite successful with rigid body docking methods. For the unbound–unbound docking, the separately crystallized component structures are used as input for docking which is more challenging than the former. Since the component structures are slightly different from the subunits in the complex structures (RMSD 0.5–1.0A).

In order to test the performance of the new docking approaches, researchers apply their docking methods to a set of complex structures in which both of bound and unbound structures are known. Chen and Weng (2003) developed a benchmark of 59 non-redundant protein complexes including 22 enzyme-inhibitor complexes, 19 antibody-antigen complexes, 11 other complexes, and 7 difficult test cases. This benchmark is widely used by other groups (Comeau et al. 2004; Duan et al. 2005; Gray 2003) to test their docking methods. Some other groups (Fernandez-Recio et al. 2002; Gabb et al. 1997; Palma et al. 2000) use their own protein complex structure data set of which some are also included in this benchmark.

A new update benchmark (Mintseris et al. 2005) was published by the same group. The new benchmark consists of 72 unbound–unbound cases, with 52 rigid-body cases, 13 medium-difficulty cases, and 7 high-difficulty cases with substantial conformational change. In addition, 12 antibody-antigen test cases are included with the antibody structure in the bound form. Yet until now, few docking groups test their docking algorithm on this new benchmark dataset.

For the evaluation a new docking algorithm, researchers usually choose complex protein data set of their own opinion or from the benchmark used by other groups, for example, the one from Chen and Weng (2003). To judge a docking algorithm is good or not, the docked complex structures are compared to the native complex structures. If the near-native structures are found in the top 100–1,000 solutions as many as possible and the RMSD value of the best hit is below 3 A, we can say this is a good docking approach. The number of hits in the top 100–1,000 docked solutions, the ranking of the best hit and the RMSD value of the best hit are the three mainly used parameters for evaluation of docking algorithms.

Moreover, investigation on the interfaces of known protein–protein complexes have revealed that enzyme-inhibitor, antibody-antigen, other complexes present important differences in the amino acid composition, hydrophobicity, and electrostatics (Decanniere et al. 2001; Glaser et al. 2001). Jackson (1999) compared protein–protein interactions in different types of complexes and concluded that enzyme-inhibitors are more static and hence more easily predictable than antibody-antigen interfaces. This suggests that different filtering criteria should be applied to different type of complexes. Li et al. applied type-dependent filtering technique to docking algorithm and retained much more native-like structures and increased the successful probability of predicting complex structures.

In the review paper of Vajda and Camacho (2004), a classification of protein complexes based on docking difficulty was introduced. They claimed that enzyme-inhibitor complexes can be determined by current docking methods with reasonable accuracy-possibly within a few alternative structures. Results for antigen–antibody pairs were less predictable, and data for small signaling complexes were generally poor. Transient complexes with large interface areas underwent substantial conformational change and were beyond the reach of current docking methods. Moreover, based on measurements of conformational change, interface area, and hydrophobicity, they defined five types of protein–protein complexes to characterize the expected level of docking difficulty.

2.4.4 Critical Assessment of Predicted Interactions

CAPRI is a community wide experiment to assess the capacity of protein-docking methods to predict protein–protein interactions (Janin et al. 2003; Mendez et al. 2003). Nineteen groups participated in rounds 1 and 2 of CAPRI and submitted blind structure predictions for seven protein–protein complexes based on the known structure of the component proteins. The predictions were compared to the unpublished X-ray structures of the complexes. CAPRI has already been a powerful drive for the community of computational biologists who developed docking algorithms. Each participating group is allowed to submit 10 models per target and these models are compared to newly obtained X-ray structures of the complexes, which crystallographers had made available for the evaluation.

The CAPRI experiments are hosted by Hendrick Kim group at the European Bioinformatics Institute (EBI). The website is http://capri.ebi.ac.uk/. In each round, one or more targets are realized and the participants have to submit their predictions before the deadline. After the submission deadline, the results will be published on the website and are classified into: "Removed predictions", "Incorrect predictions", "Acceptable predictions", "Medium predictions", and "High quality predictions", based on several criteria such as fraction of native residue contact, the RMSD values of the ligands after superimposing the receptors of the prediction and the native complex structures. Until now (December 2007), there have been 30 targets being evaluated at CAPRI experiments. These targets can be used as a benchmark data set, complementary to the Weng's docking benchmark data set.

2.4.5 Protein Interface Prediction and Protein Docking

As discussed above, although there are already many docking approaches which can predict the complex structures correctly and rank the near-native structures at the top 100 even 10, it is still very attractive to improve the current docking approaches. Improving the protein docking involves two tasks: an efficient search procedure and a good scoring function. The two critical elements in a search procedure are speed and effectiveness in covering the relevant conformational space. On the other hand, the scoring function should be fast enough to allow its application to a large number of potential solutions and, in principle, effectively discriminate between near native and non-native docked complex structures. Predicting the interface correctly first is very useful since it can help us to improve our discrimination. Therefore, the third open question in this context is 'Can we improve protein–protein docking using interface prediction?' However, although many binding sites prediction methods have been developed, only a few groups integrated it into docking. For example, predicted protein–protein binding sites first using their own prediction program: ProMate (Neuvirth et al. 2004), of which the success rate was about 70 %, and then they used these predicted binding sites to calculate the tightness of fit of the two docked proteins. A linear relation between this score and the RMSD relative to the true structure is found in most of the cases they evaluated. Gunther et al. presented ISEARCH approach which uses known domain–domain interfaces (DDI) stored in an interface library to screen unbound proteins for structurally similar interaction sites. First, a known DDI library is derived from the PDB and SCOP. Then the ligand's and the receptor's backbone structures are examined for sites similar to a representative DDI in the DDI library, using the superposition algorithm Needle Haystack (Hoppe and Frömmel 2003). When the algorithm detects local structural similarity between the receptor and one part of the DDI patch, as well as between the ligand and the corresponding patch of the DDI, a hit is obtained. Ligand and receptor are then transformed according to their superposition onto the corresponding DDI patches to build the final model of the complex. This approach was evaluated on 59 complexes from the Chen benchmark dataset (Chen and Weng 2003) and achieved acceptable docking results. These studies encourage us that using predicted interaction sites can improve protein docking. However, we will try different interface prediction methods and develop different scoring functions based on these predictions. All scoring functions will be integrated together to improve docking.

Furthermore, an important ingredient for achieving successful docking remains the use of prior knowledge of the protein regions those are likely to interact (Mendez et al. 2005). The available biochemical data (mutagenesis experiments, sequence conservation, NMR studies, etc.) relevant to the protein–protein complex may be incorporated into docking algorithms to improve efficiency. For instance, Barahona and Krippahl (2008); Krippahl and Barahona (2005) applied Constraint Programming techniques to rigid body docking algorithm BIGGER (Palma et al. 2000). Their approach imposed a broad range of constraints or combinations of constraints on distances between points of the two structures to dock, which allows the use of experimental data to increase the effectiveness and speed of modeling protein interactions and which cannot be done as efficiently in Fourier transform methods. Recently, Motiejunas et al. (2008) presented an efficient Brownian Dynamics (BD) algorithm that mimics the physical process of diffusional association and relevant biochemical data is directly incorporated as distance constraints in BD simulation.

2.5 Flexible Docking

In the previous discussion, proteins are assumed to be rigid-body. However, in native proteins, they are not rigid. The 'native state' is not a single conformation, but a whole ensemble of conformations which are populated under physiological conditions. In many cases, this flexibility is essential for the biological function. For example, it allows enzymes to fit themselves around their ligands, and molecular motors to convert chemical energy into mechanical work. Therefore, characterizing the flexibility of a native protein is of great importance in protein docking. The development of computational approaches able to accurately handle the flexibility of the protein within the context of protein–protein docking problems is still not satisfied. Modeling the flexibility of protein is still a hard problem due to high degrees of freedom and is very challenging in the next few years (Palma et al. 2000).

Conformation changes that occur during the formation of a protein complex are among the most difficult challenges to rigid body docking methods. Treating this molecule flexibility in an explicit way is an impracticable computational task since there are many degrees of freedom. Recently, many of the groups which developed docking programs have already devised a way of considering side chain conformational rearrangement during docking, whether implicitly or explicitly (Camacho 2005; Camacho and Vajda 2001; Fernendez-Recio et al. 2003; Jackson et al. 1998; Munoz et al. 2003; Palma et al. 2000; Zacharias 2003); Methods for considering higher levels of flexibility, involving the rearrangement of segments of the protein backbone, i.e. loops, domains, or the whole protein, are currently being explored by an increasing number of groups (Dominguez et al. 2003; Fitzjohn and Bates 2003; Smith et al. 2005). The limitation of current approaches in representing macromolecular flexibility for protein–protein docking have been described in a survey (Bonvin 2006) which emphasized the need for combining existing approaches.

Protein docking algorithms dealing with flexibility can be grouped into two classes: side chain flexibility only as well as backbone and side chain flexibility both considered. Molecular dynamics simulation or Monte Carlo simulation is often used in the latter approach, in conjunction with some form of rigid-body docking, either before or after the MD simulations.

2.5.1 Side-Chain Flexibility

One approach to address the problem of side-chain flexibility explicitly is to use a reduced protein model in the context of rigid-body docking, to allow some tolerance of atomic clashes across protein interfaces (Gray 2003; Zacharias 2003). Gray (2003) used a reduced representation model for side-chains in Monte Carlo search. In their low-resolution representation, each residue is represented by the four backbone atoms (N, C α, C, and O) and one pseudo-atom, the "centroid", to represent the side-chain. The location of the centroid is the average location of the side-chain atoms in the residues of identity, derived from the known structures from the PDB. After the low-resolution MC search, explicit side-chains are added to the protein backbones using a backbone-dependent rotamer packing algorithm and are optimized using a simulated-annealing Monte Carlo search. (Zacharias 2003) took a similar approach but he used at most three pseudo-atoms to represent one residue. This reduced protein representation allowed an efficient energy minimization search in rotational and translational degrees of freedom. A multi copy approach was used to select the most favorable side-chain conformation during the docking process.

Kimura et al. (2001) and Rajamani et al. (2004) have shown that the side-chains important for molecular recognition acquired conformations similar to those in which they found buried in bound complexes, using molecular dynamics. The side-chains on the interface in the unbound structures often differ a lot from those in the bound complexes (Camacho 2005). Analysis of side-chain rotamer conformations acquired in solution plays an important role in modeling the whole protein complexes. The challenging thing here is how to identify the right rotamer conformation, i.e. the rotamer conformation in the bound complex, of interface residues (if they are known to be on interface) in the unbound structure. A simple method is to use the most recurrent rotamer conformations in MD simulation snapshots (Camacho 2005). This method works quite well for Cohesin-Dockerin complex in CAPRI-II, in which the anchor residues (the most important residues in molecular recognition) do not change conformation upon binding in backbones.

Alternatively, Lorber et al. (2002) pre-calculated multiple conformations of multiple residues for the ligand. These conformations were docked into both the bound and unbound structures of the cognate receptors, and their energies were evaluated using an atomistic potential function. Their docking results from 7 test systems suggested that the pre-calculated ensembles did include side-chain conformations similar to those adopted in the experimental complexes. When docked against the bound conformations of the receptors, the near-native complexes of the unbound ligand were always distinguishable from the non-native complexes. When docked against the unbound conformations of the receptors, the near-native docking solutions can usually, but not always, be distinguished from the non-native complexes. In every case, docking the unbound ligands with flexible side chains led to better energies and a better distinction between near-native and non-native fits (Lorber et al. 2002).

2.5.2 Backbone Flexibility

Protein interfaces exhibit considerable plasticity, and various types of backbone conformational changes have been observed upon complexation. Several promising approaches have been developed to treat backbone flexibility explicitly in protein docking. In the CAPRI rounds (3–5), Smith et al. (2005) showed that using the ensembles generated by Molecular Dynamics simulations, as inputs for a rigid-body docking algorithm, increased the success rate, especially for the targets exhibiting substantial amounts of induced fit. In their recent work on CAPRI rounds, this cross-docking was followed by a short MD-based local refinement for the subset of solutions with the lowest interaction energies after minimization (Król et al. 2007). Their results revealed that cross-docking approach produces more near-native solutions but only for targets with large conformational changes upon binding. Refinement MD simulations substantially increased the fraction of native contacts for near-native solutions, but generally worsen interface and ligand RMSD. They found that although MD simulations are able to improve side-chain packing across the interface, which resulted in an increased fraction of native contacts, they are not capable of improving interface and ligand backbone RMSD for near-native structures beyond 1.5 and 3.5 Å, respectively.

Bastard et al. (2003) explored a similar approach and they took into account the induced conformational adjustment of flexible loops situated at a protein/macro-molecule interface. Their method was based on a multiple copy representation of the loops, coupled with a Monte Carlo conformational search of the relative position of the proteins and their side chain conformations. The selection of optimal loop conformations took place during Monte Carlo cycling by the iterative adjustment of the weight of each copy. Using this approach they were able to pick up the correct loop structure of a protein/DNA complex between the prd paired domain protein and it cognate DNA. HADDOCK performs rigid-body docking followed by a molecular dynamics simulated annealing refinement on backbone and side-chain degrees of freedom, and the added flexibility improves the docking results (Dominguez et al. 2003). A multi-body docking approach has been implemented in FlexDock to deal with hinge motions associated with complex formation given the knowledge of hinge regions prior to the docking and the method was able to correctly model large conformational changes occurring in the binding of calmodulin and a target peptide (Schneidman et al. 2005). Recently Wang et al. (2007) presented a reformulation of the Rosetta docking method (Gray 2003) that incorporates explicit backbone flexibility in protein–protein docking. This method was based on a "fold-tree" representation of the molecular system (Bradley and Baker 2006), which seamlessly integrates internal torsional degrees of freedom and rigid-body degrees of freedom. Problems with internal flexible regions ranging from one or more loops or hinge regions to all of one or both partners can be readily treated using appropriately constructed fold trees. The explicit treatment of backbone flexibility improves both sampling in the vicinity of the native docked conformation and the energetic discrimination between near-native and incorrect models (Wang et al. 2007).

Ehrlich et al. (2005) have studied the impact of protein flexibility on protein–protein association by means of rigid body and torsion angle dynamics simulation. In their study, the binding of barnase and barstar was chosen as a model system because the complexation of these two proteins is well-characterized experimentally and the conformational changes accompanying binding are modest. On the side chain level, they showed that the orientation of particular residues at the interface have a crucial influence on the way contacts are established during docking from short protein separations of approximately 5 Å. However, side-chain torsion angle dynamics simulations did not result in satisfactory docking of the proteins when using the unbound protein structures. On the backbone level, even small (2 Å) local loop deformations affect the dynamics of contact formation upon docking. This result indicated that both side-chain and backbone levels of flexibility influence short-range protein–protein association and should be treated simultaneously for atomic-detail computational docking of proteins.

2.5.3 Soft Docking

The center idea of the soft representation of protein surface proposed by Munoz et al. (2003) is that the assumption that a protein molecule is constituted by a hard core part that determined the overall shape of the protein surrounded by a layer of high plasticity that allows penetration of side-chains but without constraining them by type of amino acid.

Palma et al. (2000) observed that most of the conformational change upon complex formation is due to flexible amino acid side-chains positioned at the molecule surface and not every amino acid showed the same degree of freedom to move. Among the protein complexes observed, ARG, LYS, ASP, GLU, and MET presented the highest frequency and amplitude of movements between the structures of unbound and bound proteins. In their docking approach BIGGER, every atom (except C β atom) belonging to the side-chain of these five amino acids was considered flexible and was allowed to unrealistically penetrate the other molecule during the docking search.

References

Barahona P, Krippahl L (2008) Constraint programming in structural bioinformatics. Constraints J 13(1):3–20 (Constraint programming in structural bioinformatics)

Bastard K, Thureau A, Lavery R, Prevost C (2003) Docking macromolecules with flexible segments. J Comput Chem 24:1910–1920

Ben-Zeev E, Eisenstein M (2003) Weighted geometric docking: incorporating external information in the rotation-translation scan. Proteins 52:24–27

Berchmanski A, katchalski Katzir E, Eisenstein M (2002) Electrostatics in protein–protein docking. Protein Sci 11:571–587

Binkowski T, Naghibzadeh S, Liang J (2003) Castp: computed atlas of surface topography of proteins. Nucleic Acids Res 31(13):3352–3355

Bonvin A (2006) Flexible protein–protein docking. Curr Opin Structl Biol 16(2):194–200

Bowers P, Pellegrini M, Thompson M, Fierro J, Yeates T, Eisenberg D (2004) Prolinks: a database of protein functional linkages derived from coevolution. Genome Biol 5(5):R35

Bradford J, Westhead D (2005) Improved prediction of protein–protein binding sites using a support vector machines approach. Bioinformatics 21(8):1487–1494

Bradley P, Baker D (2006) Improved beta-protein structure prediction by multilevel optimization of nonlocal strand pairings and local backbone conformation. Proteins 65(4):922–929

Brady G, Stouten P (2000) Fast prediction and visualization of protein binding pockets with pass. J Comput Aided Mol Des 14:383–401

Camacho C (2005) Modeling side-chains using molecular dynamics improve recognition of binding region in capri targets. Proteins 60:245–251

Camacho C, Vajda S (2001) Protein docking along smooth association pathway. PNAS 98:10636–10641

Chen R, Weng Z (2003) A novel shape complementarirty scoring function for protein–protein docking. Proteins 51:397–408

Chen H, Zhou H-X (2005a) Prediction of interface residues in protein-protein complexes by a consensus neural network method: test against nmr data. Proteins 61(1):21–35

Chen H, Zhou H-X (2005b) Prediction of interface residues in protein-protein complexes by a consensus neural network method: test against nmr data. Proteins 61(1):21–35

Chothia C, Janin J (1975) Principles of protein–protein recognition. Nature 256(5520):705–708

Comeau S, Gatchell D, Vajda S, Camacho C (2004) Cluspro: an automated docking and discrimination method for the prediction of protein complexes. Bioinformatics 20:45–50

Criekinge WV, Beyaert R (1999) Yeast two-hybrid: State of the art. Biol Proced Online 2:1–38

Decanniere K, Transue T, Desmyter A, Maes D, Muyldermans S, Wyns L (2001) Degenerate interfaces in antigen-antibody complexes. J Mol Biol 313:473–478

Dominguez C, Boelens R, Bonvin A (2003) Haddock: a protein–protein docking approach based on biochemical and/or biophysical information. J Am Chem Soc 125:1731–1737

Duan Y, Reddy V, Kaznessis Y (2005) Physicochemical and residue conservation calculations to improve the ranking of protein–protein docking solutions. Protein Sci 14:316–328

Edelsbrunner H, Facello M, Liang Jie (1998) On the definition and the construction of pockets in macromolecules discrete. Appl Math 88:83–102

Ehrlich L, Nilges M, Wade R (2005) The impact of protein flexibility on protein-protein docking. Proteins 58(1):126–133

Eisenstein z, Katchalski-Katzir E (2004) On proteins, grids, correlations, and docking. C.R Biol 327:409–420

Fernandez-Recio J, Totrov M, Abagyan R (2002) Soft protein–protein docking in internal coordinates. Protein Sci 11:280–291

Fernandez-Recio J, Totrov M, Abagyan R (2004) Identification of protein-protein interaction sites form docking energy landscapes. J Mol Biol 335:843–865

Fernendez-Recio J, Totrov M, Abagyan R (2003) Icm-disco docking by global energy optimization with fully flexible side-chains. Proteins 52:113–117

Fitzjohn P, Bates P (2003) Guided docking: first step to locate potential binding sites. Proteins 52:28–32

Gabb H, Jackson R, Sternberg M (1997) Modelling protein docking using shape complimentarity, electrostatics and biochemical information. J Mol Biol 272(1):106–120

Galperin M, Koonin E (2000) Who's your neighbor? new computational approaches for functional genomics. Nat Biotechnol 18(6):609–613

Glaser F, Steinberg D, Vakser I, Ben-Tal N (2001) Residue frequencies and pairing preferences at protein–protein interfaces. Proteins 43:82–102

Glaser F, Morris R, Najmanovich R, Laskowski R, Thornton J (2006) A method for localizing ligand binding pockets in protein structures. Proteins 62:479–488

Goh C, Bogan A, Joachimiak M, Walther D, Cohen F (2000) Co-evolution of proteins with their interaction partners. J Mol Biol 299(2):283–293

Gray J, Moughon S, Wang C, Schueler-Furman O, Kuhlman B, Rohl C, Baker D (2003) Protein-protein docking with simultaneous optimization of rigid-body displacement and side-chain conformations. J Mol Biol 331:281–299

Hadley C, Jones DT (1999) A systematic comparison of protein structure classifications: SCOP, CATH and FSSP. Structure 7(9):1099–1112

Halperin I, Ma B, Wolfson H, Nussinov R (2002a) Principles of docking: an overview of search algorithms and a guide to scoring functions. Proteins 47:409–443

Halperin I, Ma B, Wolfson H, Nussinov R (2002b) Principles of docking: an overview of search algorithms and a guide to scoring functions. Proteins 47:409–443

Hendlich MF, Barnickel G, Rippmann F (1997) Ligsite: automatic and efficient detection of potential small molecule-binding sites in proteins. J Mol Graph 15(6):359–363

Hoppe A, Frommel C (2003) Needlehaystack: a program for the rapid recognition of local structures in large sets of atomic coordinates. J Appl Crystallogr 36(4):1090–1097

Huynen M, Snel B, Lathe W, Bork P (2000) Predicting protein function by genomic context: quantitative evaluation and qualitative inferences. Genome Res 10(8):1204–1210

Jackson R (1999) Comparison of protein-protein interactions in serine protease-inhibitor and antibody-antigen complexes: implications for the protein docking problem. Protein Sci 8:603–613

Jackson R, Gabb H, Sternbergm M (1998) Rapid refinement of protein interfaces incorporating solvation: application to the docking problem. J Mol Biol 276:265–285

Janin J, Henrick K, Moult J, Eyck L, Sternberg M, Vajda S, Vakser I, Wodak S (2003) Capri: a critical assessment of predicted interactions. Proteins 52:2–9

Jones S, Thornton J (1997) Prediction of protein–protein interaction sites using patches analysis. J Mol Biol 272:133–143

Jothi R, Cherukuri P, Tasneem A, Przytycka T (2006) Co-evolutionary analysis of domains in interacting proteins reveals insights into domain–domain interactions mediating protein–protein interactions. J Mol Biol 362(4):861–875

Katchalski-Katzir EI, Shariv M. Eisenstein A, Friesem C, Aflalo, Vakser I (1992) Molecular surface recognition: determination of geometric fit between proteins and their ligands by correlation techniques. PNAS 89:2195–3199

Kim W, Bolser D, Park J (2004) Large-scale co-evolution analysis of protein structural interlogues using the global protein structural interactome map (psimap). Bioinformatics 20(7):1138–1150

Kimura S, Brower R, Vajda S, Camacho C (2001) Dynamical view of the positions of key side chains in protein-protein recognition. Biophysical 80(2):635–642

Koike A, Takagi T (2004) Prediction of protein–protein interaction sites using support vector machines. Protein Eng 17(2):165–173

Król M, Chaleil R, Tournier A, Bates P (2007) Implicit flexibility in protein docking: cross-docking and local refinement. Proteins 69(4):750–757

Krippahl L, Barahona P (2005) Applying constraint programming to rigid body protein docking principles and practice of constraint programming. In: Peter van Beek (ed) CP'2005 (Procs). Lecture Notes in Computer Science, vol 3709. Springer, pp 373–387

Laskowski R (1995) Surfnet: a program for visualizing molecular surfaces, cavities and intermolecular interactions. J Mol Graph 13:323–330

Laskowski RA, Rullmannn JA, MacArthur MW, Kaptein R, Thornton JM (1996) AQUA and PROCHECK-NMR: programs for checking the quality of protein structures solved by NMR. J Biomol NMR 8:477–486

Laurie A, Jackson R (2005) Q-sitefinder: an energy-based method for the prediction of protein-ligand binding sites. Bioinformatics 21:1908–1916

Lee B, Richards FM (1971) The interpretation of protein structures: estimation of static accessibility. J Mol Biol 55(3):379–400

Levitt D, Banaszak L (1992) Pocket: a computer graphics method for identifying and displaying protein cavities and their surrounding amino acids. J Mol Graph 10:229–234

Liang J, Edelsbrunner H, Woodward C (1998) Anatomy of protein pockets and cavities: measurement of binding site geometry and implications for ligand design. Protein Sci 7:1884–1897

Liang S, Zhang J, Zhang S, Guo H (2004) Prediction of the interaction site on the surface of an isolated protein structure by analysis of side chain energy scores. Proteins 57:548–557

Liang S, Zhang C, Liu S, Zhou Y (2006) Protein binding site prediction using an empirical scoring function. Nucleic Acids Res 34(13):3698–3707

Lorber D, Udo M, Shoichet B (2002) Protein-protein docking with multiple residue conformations and residue substitutions. Protein Sci 11:1393–1408

Marcotte C, Marcotte E (2002) Predicting functional linkages from gene fusions with confidence. Appl Bioinform 1(2):93–100

Marcotte E, Pellegrini M, Ng H, Rice D, Yeates T, Eisenberg D (1999) Detecting protein function and protein–protein interactions from genome sequences. Science 285(5428):751–753

Mendez R, Leplae R, Maria L, Wodak S (2003) Assessment of blind predictions of protein–protein interactions: current status of docking methods. Proteins 52:51–67

Mendez R, Leplae R, Lensink MF, Wodak SJ (2005) Assessment of capri predictions in rounds 3–5 shows progress in docking procedures. Proteins 60(2):150–169

Mintseris J, Wiehe K, Pierce B, Anderson R, Chen R, Janin J, Weng Z (2005) Protein–protein docking Benchmark 2.0: an update. Proteins 60(2):214–216

Mirny L, Shakhnovich E (1999) Universally conserved positions in protein folds: reading evolutionary signals about stability, folding kinetics and function. J Mol Biol 291(1):177–196

Morris G, Goodsell D, Halliday R, Huey R, Hart W, Belew R, Olson A (1998) Automated docking using a lamarckian genetic algorithm and an empirical binding free energy function. J Comput Chem 19:1639–1662

Motiejunas D, Gabdoulline RR, Wang T, Feldman-Salit A, Johann T, Winn PJ, Wade RC (2008a) Protein-protein docking by simulating the process of association subject to biochemical constraints. Proteins 33:125

Motiejunas D, Gabdoulline RR, Wang T, Feldman-Salit A, Johann T, Winn PJ, Wade RC (2008b) Protein-protein docking by simulating the process of association subject to biochemical constraints. Proteins 33:125

Munoz C, Peissker T, Yoshimori A, Ichiish E (2003) Docking unbound proteins with miax: a novel algorithm for protein–protein soft docking. Genome Informationics 14:238–249

Murzin A, Brenner S, Hubbard T, Chothia C (1995) Scop: a structural classification of proteins database for the investigation of sequences and structures. J Mol Biol 247:536–540

Neuvirth H, Raz R, Schreiber G (2004) Promate: a structure based prediction program to indentify the location of protein–protein binding stes. J Mol Biol 338:181–199

Orengo CA, Michie AD, Jones S, Jones DT, Swindells MB, Thornton JM (1997) CATH–a hierarchic classification of protein domain structures. Structure 5(8):1093–1108

Overbeek R, Fonstein M, D'Souza M, Pusch G, Maltsev N (1999) The use of gene clusters to infer functional coupling. Proc Natl Acad SciU S A 96(6):2896–2901

Palma P, Krippahl L, Wampler J, Moura J (2000) Bigger: a new (soft) docking algorithm for predicting protein interactions. Proteins 39(4):372–384

Pellegrini M, Marcotte E, Thompson M, Eisenberg D, Yeates T (1999) Assigning protein functions by comparative genome analysis: protein phylogenetic profiles. Proc Natl Acad SciU S A 96(8):4285–4288

Phizicky EM, Fields S (1995) Protein–protein interactions: methods for detection and analysis. Microbiol Rev 59(1):94–123

Porollo A, Meller J (2007) Prediction-based fingerprints of protein–protein interactions. Proteins 66(3):630–645

Rajamani D, Thiel S, Vajda S, Camacho C (2004) Anchor residues in protein-protein interactions. PNAS 101:11287–11292

Recio J, Totrov M, Skorodumov C, Abagyan R (2005) Optimal dockingarea: a new method for predicting protein–protein interaction sites. Proteins 58:134–143

Rogozin I, Makarova K, Murvai J, Czabarka E, Wolf Y, Tatusov R, Szekely L, Koonin E (2002) Connected gene neighborhoods in prokaryotic genomes. Nucleic Acids Res 30(10):2212–2223

Schneidman D, Inbar Y, Nussinov R, Wolfson H (2005) Geometry-based flexible and symmetric protein docking. Proteins 60(2):224–231

Sen T, Kloczkowski A, Jernigan R, Yan C, Honavar V, Ho K, Wang CZ, Ihm Y, Cao H, Gu X, Dobbs D (2004) Predicting binding sites of hydrolase-inhibitor complexes by combining several methods. BMC Bioinform 5:205

Shoemaker B, Panchenko A (2007) Deciphering protein–protein interactions. part II. computational methods to predict protein and domain interaction partners. PLoS Comput Biol 3(4):e43

Smith G, Sternberg M, Bates P (2005) The relationship between the flexibility of proteins and their conformational states on forming protein–protein complexes with an application to protein–protein docking. J Mol Biol 345:1077–1101

Snitkin E, Gustafson A, Mellor J, Wu J, DeLisi C (2006) Comparative assessment of performance and genome dependence among phylogenetic profiling methods. BMC Bioinform 7:420

Sternberg M, Gabb H, Jackson R (1998) Predictive docking of protein–protein and protein-dna complexes. Curr Opin Struct Biol 8(2):265–269

Tovchigrechko A, Wells C, Vakser I (2002) Docking of protein models. Protein Sci 11:1888–1896

Vajda S, Camacho C (2004) Protein-protein docking: is the glass half-full or half-empty? Trends Biotechnol 22(3):110–116

Wang C, Bradley P, Baker D (2007) Protein-protein docking with backbone flexibility. J Mol Biol 373(2):503–519

Yan C, Dobbs D, Honavar V (2004) A two stage classifier for identificantion protein–protein interface residue. Bioinformatics 20:i371–i378

Yanai I, Derti A, DeLisi C (2001) Genes linked by fusion events are generally of the same functional category: a systematic analysis of 30 microbial genomes. Proc Natl Acad Sci U S A 98(14):7940–7945

Young KH (1998) Yeast two-hybrid: so many interactions, (in) so little time. Biol Reprod 58(3):302–311

Zacharias M (2003) Protein–protein docking with a reduced protein model accounting for side-chain flexibility. Protein Sci 12:1271–1282

Zhou H, Shan Y (2001) Prediction of protein interaction sites from sequence profile and residue neighbor list. Proteins 44(3):336–343

Chapter 3
Materials and Methods

3.1 Selection of Sequences, 3D Structures and Multiple Sequence Alignment

Sequences used in the study were obtained from NCBI (www.ncbi.nlm.nih.gov) and 3D structures viz. endo-1,4-beta xylanases (1YNA) were obtained from RSCB (http://www.pdb.org/pdb/home/home.do). Further conserved regions were predicted by performing multiple sequence alignment using ClustalX 2.0.10 software (Larkin et al. 2007).

3.2 Mutant Generation

Mutation in endo-1, 4-beta-xylanase (1YNA) was carried out using Swiss PDB-Viewer (Guex and Peitsch 1997) and mutant was validated by PROCHECK (http://www.ebi.ac.uk/thornton-srv/software), ERRAT(http://nihserver.mbi.ucla.edu).

3.3 Energy Minimization

An energy minimization method was carried out using Swiss-PDB viewer which includes a version of the GROMOS 43B1 force field which allows evaluating the energy of a structure as well as repairing distorted geometries through energy minimization in vacuo without any reaction field (Christen et al. 2005). After minimization of the structure, the one with lower energy was validated using PROCHECK (Laskowski et al. 1993, 1996) and ERRAT (Colovos and Yeates 1993) available at structural analysis and validation server (http://nihserver.mbi.ucla.edu/SAVES).

M. V. K. Karthik and P. Shukla, *Computational Strategies Towards Improved Protein Function Prophecy of Xylanases from* Thermomyces lanuginosus, SpringerBriefs in Systems Biology, DOI: 10.1007/978-1-4614-4723-8_3, © The Author(s) 2012

Table 3.1 BLOCKS and their position

BLOCKS	Position (score)	Description	Related sequence
IPB001137C	70...98(1604)	Glycoside hydrolase, family 11	209
IPB001137D	106...156(1483) 105...155(1015)	Glycoside hydrolase, family 11	209
IPB001137A	8...21(1300)	Glycoside hydrolase, family 11	209
IPB001137B	32...51(1288)	Glycoside hydrolase, family 11	209

3.4 Stereo-Chemical Quality Check and Analysis of Non-Bonded Interactions

PROCHECK was used to evaluate the stereo chemical quality of all 19 mutant generated through Swiss PDB-Viewer after permutation analysis. Further ERRAT was used to analyze the statistics of non-bonded interactions between different atom types. Further plots were generated detailing the value of the error function versus position of a 9-residue sliding window and were designed from highly refined structures.

3.5 Protein–Protein Docking

Hex 5.1 docking program was used for protein molecules docking calculations (Ritchie and Venkatraman 2010). In these calculations, each molecule corresponds to 3D parametric functions describing surface shape, electrostatic charge, and potential distributions. In this study, we used substrates viz. xylobiose and beta-D-xylopyranose as a ligand and these were docked with modeled 19 mutants (Table 3.1) and 1YNA wild type where amino acids involved in interactions were identified.

Further, electrostatic and van-der-Waals interactions were also taken into account in our calculations. The expression for docking score was derived as function of the six degrees of freedom in rigid body docking search by accession of the mutual overlapping score for parametric functions.

3.6 InterProScan

InterProScan from EBI's ftp server (ftp://ftp.ebi.ac.uk/pub/databases/interpro/iprscan) (Zdobnov and Apweiler 2001) was used for identifying motifs of family 11 xylanases as databases of protein domains and functional sites are vital possessions for the prediction of protein functions. This database integrates PROSITE, PRINTS, ProDom, SMART, and TIGRFAMs databases providing required accuracy.

3.7 Identification of Functionally Important Regions of Wild Type viz. Endo-1, 4-beta xylanases

SiteID^{TM} is used to identify and visualize protein binding sites. ConSurf server is a useful and user-friendly tool that enables the identification of functionally important regions on the surface of a protein or domain, of known 1YNA three-dimensional (3D) structure, based on the phylogenetic relations between its close sequences homologues.

3.8 3d2GO Server: From Protein 3D Structure to Gene Ontology Term

3d2GO (http://www.sbg.bio.ic.ac.uk/phyre/pfd/index.html) was used for methods of function prediction, using sequence and structure, to predict Gene Ontology (GO) terms for protein (Tung et al. 2007; Ortiz et al. 2002; Altschul et al. 1997; Edgar and Robert 2004; Capra and Singh 2007; Edelsbrunner et al. 1998; Moll and Kavraki 2008; Hawkins et al. 2006). The mutant PDB structure of 19 mutants as stated in Table 3.1 were sent to 3d2GO server and results were retrieved. The Support Vector Machine (SVM) learning algorithm was constituted and a benchmark of diverse proteins with known GO annotations thus discriminating true and false positive annotations.

References

Larkin MA, Blackshields G, Brown NP (2007) Clustal W and clustal X version 2.0. Bioinformatics 23:2947–2948

Guex N, Peitsch MC (1997) Swiss-model and the swiss-Pdbviewer: an environment for comparative protein modelling. Electrophoresis 18:2714–2723

Christen M, Hünenberger PH, Bakowies D, Baron R, Bürgi R, Geerke DP, Heinz TN, Kastenholz MA, Kräutler V, Oostenbrink C, Peter C, Trzesniak D, van Gunsteren WF (2005) The GROMOS software for biomolecular simulation: GROMOS05. J Comput Chem 26:1719–1751

Laskowski RA, MacArthur MW, Moss DS, Thornton JM (1993) PROCHECK—a program to check the stereochemical quality of protein structures. J App Cryst 26:283–291

Laskowski RA, Rullmannn JA, MacArthur MW, Kaptein R, Thornton JM (1996) AQUA and PROCHECK-NMR: programs for checking the quality of protein structures solved by NMR. J Biomol NMR 8:477–486

Colovos C, Yeates TO (1993) Verification of protein structures: patterns of non-bonded atomic interactions. Prot Sci 9:1511–1519

Ritchie DW, Venkatraman V (2010) Ultra-fast FFT protein docking on graphics processors. Bioinformatics 26:2398–2405

Zdobnov EM, Apweiler R (2001) InterProScan—an integration platform for the signature-recognition methods in InterPro. Bioinformatics 17:847–848

Tung CH, Huang JW, Yang JM (2007) Kappa-alpha plot derived structural alphabet and BLOSUM-like substitution matrix for fast protein structure database search. Genome Biol 8:31.1–31.16

Ortiz AR, Strauss CE, Olmea O (2002) Mammoth (matching molecular models obtained from theory): an automated method for model comparison. Prot Sci 11:2606–2621

Altschul SF, Madden TL, Schaffer AA, Zhang J, Zhang Z, Miller W, Lipman DJ (1997) Gapped BLAST and PSI-BLAST: a new generation of protein database search programs. Nucleic Acids Res 25:3389–3402

Edgar, Robert C (2004) MUSCLE: multiple sequence alignment with high accuracy and high throughput. Nucleic Acids Res 32:1792–1797

Capra J, Singh M (2007) Predicting functionally important residues from sequence conservation. Bioinformatics 23:1875

Edelsbrunner H, Facello M, Jie Liang (1998) On the definition and the construction of pockets in macromolecules.Discrete Appl Math 88:83–102

Moll M and Kavraki LE (2008) Matching of structural motifs using hashing on residue labels and geometric filtering for protein function prediction. The Seventh Annual International Conference on Computational Systems Bioinformatics, Stanford, CA

Hawkins T, Luban S, Kihara D (2006) Enhanced automated function prediction using distantly related sequences and contextual association by PFP. Prot Sci 15:1550–1556

Chapter 4
Results and Discussions

The identification of a good binding site and characterization of a target protein is a prime importance that leads to its functional annotations. At the outset 37 sequences of various family 11 xylanases were selected for multiple sequence alignment as shown in Fig. 4.1 and found Tyr77 (Y77) to be conserved.

Possible binding sites are obtained by using SYBYL Site ID module which is shown in Fig. 4.2 and found green region to be highly conserved. Predicted binding site residues are Trp18, Asn44, Val46, Tyr73, Tyr77, Ala176, and Glu178. During identification of functionally important regions through ConSurf server it was noticed that Tyr77 was conserved and it is marked in dark pink showing the representation of conserved residue in Fig. 4.3. Mutual interactions between proteins and peptides, nucleic acids, or ligands play a vital role in every biological process. Thus, a detailed understanding of the mechanism of these processes requires the identification of functionally important amino acids at the protein surface that are responsible for these interactions.

Moreover the results for InterProScan were indicated in Table 4.1 showing Tyr77 residue in glycoside hydrolase family 11 motifs with a score of 1604. By using InterProScan we also found important residues viz. blocks, motifs which are represented in Tables 4.2 and 4.3.

Molecular docking with endo-1, 4-β-xylanase (1YNA) and ligands, such as xylobiose, beta-D-xylopyranose shown Tyr77 as a common binding site residue and can be observed in Figs. 4.4 and 4.5 (Table 4.4).

From above inclusive studies we elucidated Tyr77 residue to be conserved and so rightfully taken as a target for generation of mutation through permutation method. Therefore, 19 mutations where generated through Swiss PDB viewer. Further, these mutants were stabilized by minimizing their energy. Scores of energy minimizations were presented in Table 4.5.

Stereochemical quality check and analysis of non-bonded interactions was verified by Ramachandran plot through PROCHECK, where it was noticed that the accuracy of mutant models was satisfactory as 84.7 % amino acids were found in

M. V. K. Karthik and P. Shukla, *Computational Strategies Towards Improved Protein Function Prophecy of Xylanases from* Thermomyces lanuginosus, SpringerBriefs in Systems Biology, DOI: 10.1007/978-1-4614-4723-8_4, © The Author(s) 2012

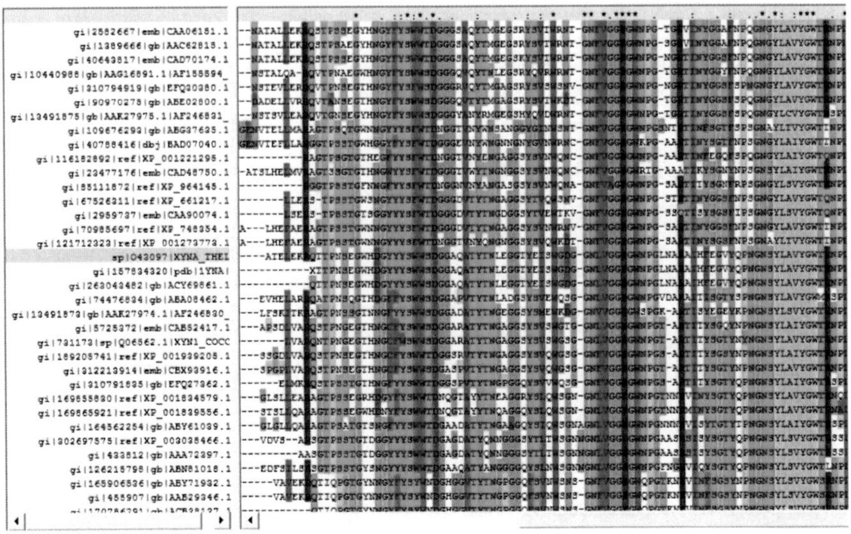

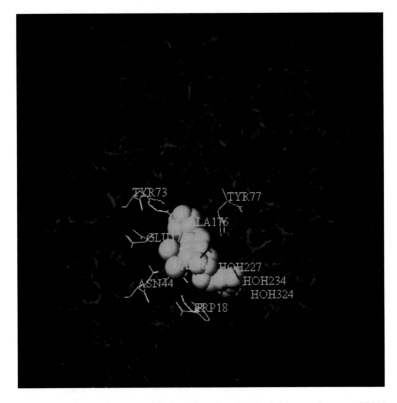

Fig. 4.1 ClustalX result showing an alignment of the amino acid sequences. *Gray shading* represents conserved amino acids which are identical in all or similar in greater than 75 % of the sequences shown, respectively

Fig. 4.2 Representation of amino acids in active site of endo-1,4-beta-xylanase (1YNA)

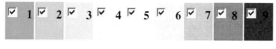

Variable   Average

Conserved

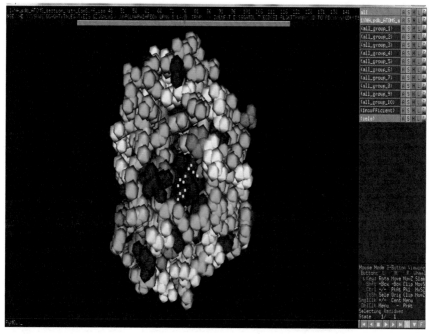

Fig. 4.3 ConSurf: server for the identification of functional regions in proteins. The conservation pattern obtained using ConSurf where conserved Tyr77 is pointed by using PYMOL

most favored region where as 15.3 % were present in additionally allowed regions and no residue was found in disallowed region. Furthermore, ERRAT studies concluded the accuracy of 89.13 % (Table 4.6).

These 19 mutants were further docked with xylobiose and beta-D-xylopyranose where we obtained three mutants viz. Y77F, Y77R, and Y77W having their mutant residue (viz. Phenylalanine, Arginine, and Tryptophan) in binding site (Figs. 4.6–4.11). Additionally docking scores for Y77W (-272.50, -171.47) and Y77R (-272.65, -174.71) for xylobiose and beta-D-xylopyranose indicating their higher stability against wild type (-271.77, -170.78). From Table 4.7 docking scores of 19 mutants were observed.

Table 4.1 InterProScan showing the motif region

Blocks Name:IPB001137C; Description: Family 11 glycoside

hydrolase

Appearance (Score):

Position	Sequence	Score
70..98	GNSYLAVYGWTRNPLVEYYIVENFGTYDP	1604

Sequence:

ETTPNSEGWHDGYYYSWWSDGGAQATYTNLEGGTYEISWG

DGGNLVGGKGWNPGLNARAI

HFEGVYQPNGNSYLAVYGWTRNPLVEYYIVENFGTYDPSSG

ATDLGTVECDGSIYRLGKT

TRVNAPSIDGTQTFDQYWSVRQDKRTSGTVQTGCHFDAWAR

AGLNVNGDHYYQIVATEGY

FSSGYARITVADVG

Red colour representing the motif of Family 11 glycoside hydrolase

Table 4.2 Prints and amino acid location

Method	AccNumber	ShortName	Location
FPrintScan	PR00911	GLHYDRLASE11	**T[73–82] 5.8e-17**
			T[83–93] 5.8e-17
			T[111–117] 5.8e-17
			T[131–136] 5.8e-17
			T[136–145] 5.8e-17
			T[159–167] 5.8e-17
HMMPfam	PF00457	Glyco_hydro_11	T[9–190] 1.4e-76

Table 4.3 BLOCKS and their position

BLOCKS	Position (score)	Description	Related sequence
IPB001137C	70..98(1604)	Glycoside hydrolase, family 11	209
IPB001137D	106..156(1483)	Glycoside hydrolase, family 11	209
	105..155(1015)		
IPB001137A	8..21(1300)	Glycoside hydrolase, family 11	209
IPB001137B	32..51(1288)	Glycoside hydrolase, family 11	209

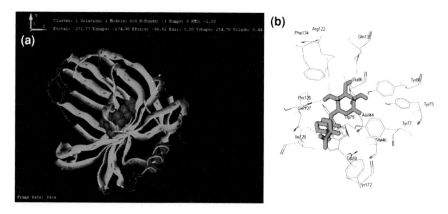

Fig. 4.4 Cartoon representation of the docked figure. **a** Docking with xylobiose. **b** Systematic representation of binding site of 1YNA and xylobiose

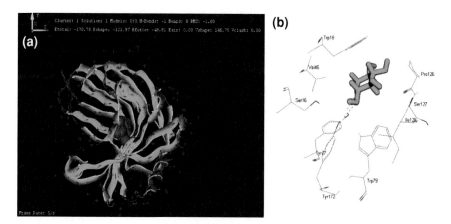

Fig. 4.5 Cartoon representation of the docked figure. **a** Docking with beta-D-xylopyranose. **b** Systematic representation of binding site of 1YNA and beta-D-xylopyranose

Table 4.4 Interactions of endo-1,4-beta-xylanase(1YNA) with xylobiose and beta-D-xylopyranose

Receptor	Docking with xylobiose (kj/mole)			Docking with 1beta-D-xylopyranose (kj/mole)		
	E shape	E force	E total	E shape	E force	E total
1YNA (wild type)	−174.96	−96.81	−271.77	−121.97	−48.81	−170.78

It was obvious that our studies are to be directed to reveal protein 3D structure to gene ontology (GO) terms which was managed through 3d2GO server. From these results we elucidated Y77W mutant is having enhanced functions

Table 4.5 Results of energy minimization of mutants

Mutants	Energy minimization						
	Bonds	Angles	Torsion	Improper	Nonbonded	Electrostatic	Total
Y77A	109.225	500.133	967.054	170.005	−6633.08	−5924.47	−10720.261
Y77C	109.126	600.461	958.068	170.136	−6650.30	−5924.95	−10737.461
Y77D	107.906	502.013	969.326	172.102	−6553.06	−692619	−10736.891
Y77E	108.700	502.340	959.268	170.056	−6559.72	−5921.99	−10741.355
Y77F	104.617	510.651	956.982	178.281	−6601.96	−5944.29	−10795.712
Y77G	109.227	500.161	957.581	170.248	−6517.33	−5852.78	−10632.894
Y77H	108.847	606.408	957.812	170.990	−6602.90	−5952.86	−10817.704
Y77I	104.890	608.043	963.340	171.632	−6559.50	−5940.24	−10761.938
Y77K	108.971	603.260	968.941	169.466	−6541.64	−6928.01	−10729.017
Y77L	104.383	616.136	963.393	172.733	−6541.42	−5942.83	−10738.603
Y77M	211.293	546.621	962.683	172.847	−6695.11	−6962.71	−10666.373
Y77N	109.362	504.011	959.431	171.147	−6557.35	−6093.43	−10906.821
Y77P	106.632	533.066	966.929	187.499	−6519.41	−5919.24	−10644.534
Y77Q	105.021	507.944	956.320	170.846	−6581.82	−6109.53	−10951.221
Y77R	108.955	502.581	960.154	170.395	−6576.70	−6188.99	−11023.614
Y77S	109.150	500.562	960.324	170.160	−6535.44	−5935.72	−10730.964
Y77T	108.030	500.002	960.209	170.042	−6551.40	−5934.05	−10744.760
Y77V	108.937	502.313	967.309	171.159	−6648.61	−6923.79	−10732.583
Y77W	122.961	549.545	963.663	186.353	−6501.72	−5933.35	−10613.558

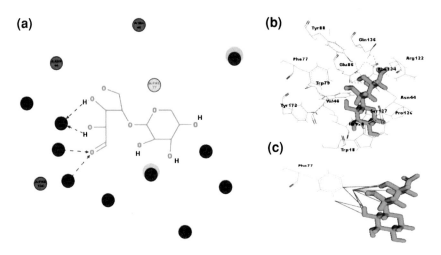

Fig. 4.6 Representation of mutant Y77F (Tyrosine (Tyr) → Phenylalanine (Phe)acid) interaction with xylobiose. **a** 2D representation of binding site where mutant is represented in *yellow color*. **b** 3D representation of binding site. **c** Mutant residue interacting with xylan

Table 4.6 Validation of mutants by using ERRAT and PROCHECK

Mutants	Errat	Ramachandran plot	
		Most favored region (%)	Additional allowed region (%)
Y77A	87.5	84.7	15.3
Y77C	87.5	84.7	15.3
Y77D	88.043	84.7	15.3
Y77E	86.957	84.7	15.3
Y77F	88.043	84.7	15.3
Y77G	86.957	84.6	15.4
Y77H	89.130	84.7	15.3
Y77I	88.043	84.7	15.3
Y77K	89.130	84.7	15.3
Y77L	87.5	84.7	15.3
Y77M	88.043	84.7	15.3
Y77N	88.043	84.7	15.3
Y77P	88.043	84.6	15.4
Y77Q	88.043	84.7	15.3
Y77R	88.043	84.7	15.3
Y77S	86.957	84.7	15.3
Y77T	86.957	84.7	15.3
Y77V	86.957	84.7	15.3
Y77W	89.130	84.7	15.3

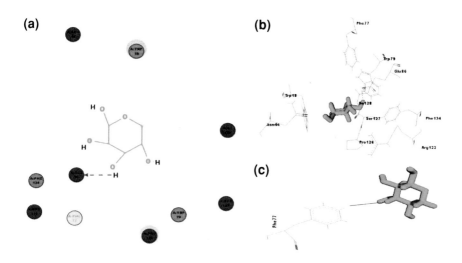

Fig. 4.7 Representation of mutant Y77F (Tyrosine (Tyr) → Phenylalanine (Phe)acid) interaction with 1beta-D-xylopyranose. **a** 2D representation of binding site where mutant is represented in *yellow color*. **b** 3D representation of binding site. **c** Mutant residue interacting with xylan

Table 4.7 Interactions of model mutants with xylobiose and beta-D-xylopyranose

Mutants	Docking with xylobiose (kj/mole)			Docking with 1beta-D-xylopyranose (kj/mole)		
	E shape	E force	E total	E shape	E force	E total
Y77A	−190.94	−69.21	−260.15	−123.62	−41.15	−164.77
Y77C	−205.01	−58.71	−263.71	−124.41	−42.20	−166.60
Y77D	−189.73	−75.23	−264.96	−126.13	−46.26	−172.38
Y77E	−190.00	−77.20	−267.19	−126.96	−42.58	−169.55
Y77F	**−205.81**	**−60.74**	**−265.81**	**−129.48**	**−41.15**	**−170.63**
Y77G	−191.50	−71.54	−263.04	−123.63	−43.93	−167.56
Y77H	−182.94	−80.40	−263.34	−114.28	−57.04	−171.33
Y77I	−206.63	−57.41	−264.03	−127.97	−37.87	−165.84
Y77K	−189.03	−76.78	−265.81	−121.18	−43.41	−164.58
Y77L	−183.72	−77.72	−261.44	−122.25	−45.36	−167.61
Y77M	−192.33	−69.59	−261.93	−118.52	−50.36	−168.88
Y77N	−184.09	−79.77	−263.86	−128.94	−45.14	−174.08
Y77P	−178.08	−87.11	−265.19	−121.00	−42.13	−163.13
Y77Q	−186.92	−73.47	−260.39	−119.92	−44.80	−164.72
Y77R	**−206.55**	**−66.10**	**−272.65**	**−134.83**	**−39.89**	**−174.71**
Y77S	−206.94	−58.21	−264.94	−124.16	−44.36	−168.53
Y77T	−205.54	−59.07	−264.61	−124.41	−42.18	−166.59
Y77V	−206.52	−58.64	−265.16	−123.40	−42.08	−165.48
Y77W	**−192.95**	**−79.55**	**−272.50**	**−138.51**	**−32.96**	**−171.47**

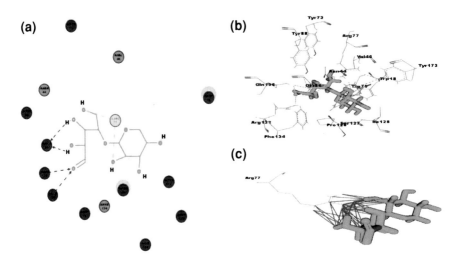

Fig. 4.8 Representation of mutant Y77R (Tyrosine (Tyr) → Arginine (Arg)) interaction with xylobiose. **a** 2D representation of binding site where mutant is represented in *yellow color*. **b** 3D representation of binding site. **c** Mutant residue interacting with xylan

(as confidence levels) on hydrolase activity (0.98), molecular function (0.58), proteolysis (0.19) and binding activity (0.17), and Y77F mutant showed good

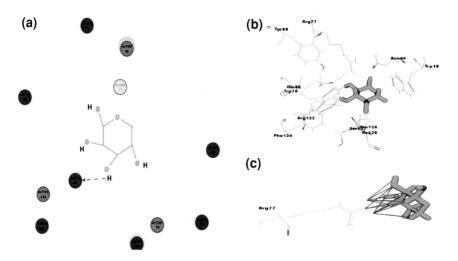

Fig. 4.9 Representation of mutant Y77R (Tyrosine (Tyr) → Arginine (Arg)) interaction with 1beta-D-xylopyranose. **a** 2D representation of binding site where mutant is represented in *yellow color*. **b** 3D representation of binding site. **c** Mutant residue interacting with xylan

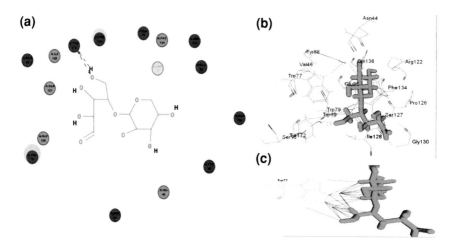

Fig. 4.10 Representation of mutant Y77W (Tyrosine (Tyr) → Tryptophan (Trp)) interaction with xylobiose. **a** 2D representation of binding site where mutant is represented in *yellow color*. **b** 3D representation of binding site. **c** Mutant residue interacting with xylan

Table 4.8 Function prediction using 3d2GO: predicted functions for protein

GO term	Description	Confidence			
		1YNA	Y77W	Y77R	Y77F
GO:0004553	Hydrolase activity, hydrolyzing O-glycosyl compounds	0.99	0.99	0.99	0.99
GO:0016798	Hydrolase activity, acting on glycosyl bonds	0.98	0.98	0.98	0.98
GO:0008152	Metabolic process	0.98	0.98	0.98	0.98
GO:0005975	Carbohydrate metabolic process	0.98	0.98	0.98	0.98
GO:0016787	Hydrolase activity	0.97	**0.98**	**0.98**	**0.98**
GO:0003673	Gene_ontology	0.46	**0.58**	**0.52**	**0.51**
GO:0003674	Molecular_function	0.46	**0.58**	**0.52**	**0.51**
GO:0003824	Catalytic activity	0.18	**0.24**	**0.20**	**0.95**
GO:0006508	Proteolysis	0.18	**0.19**	**0.19**	**0.18**
GO:0005488	Binding	0.13	**0.17**	**0.17**	**0.16**

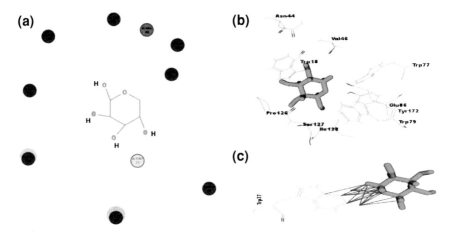

Fig. 4.11 Representation of mutant Y77W (Tyrosine (Tyr) → Tryptophan (Trp)) interaction with 1beta-D-xylopyranose. **a** 2D representation of binding site where mutant is represented in *yellow color*. **b** 3D representation of binding site. **c** Mutant residue interacting with xylan

enhanced catalytic activity with confidence level of 0.95 (Table 4.8). A graph had been plotted taking function on x-axis and confidence value on y-axis for mutants Y77W, Y77R, and Y77F (Fig. 4.12).

Confidence

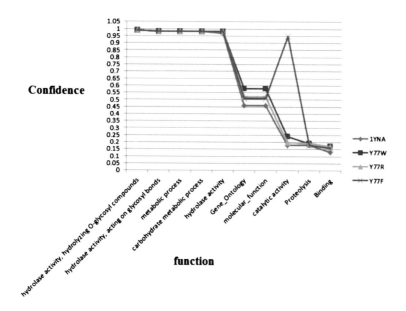

function

Fig. 4.12 Graph showing confidence versus function for 1YNA (wild type), Y77W, Y77R, and Y77F

Directed evolution has been proved to be a very powerful tool in protein engineering. A great deal of enzymes has been engineered by this approach. However, most of these mutant xylanases were engineered to be more enhanced in various aspects than the wild type.

Chapter 5
Conclusion

The identification of a high-quality binding site and depiction of a target protein is of prime significance that leads to its functional annotations. In the present work, diverse binding site residues viz. Trp18, Asn44, Val46, Tyr73, Tyr77, Ala176, Glu178 were studied using SYBYL and it was envisaged that Tyr77 was conserved in family 11 xylanases. Furthermore, during residue conservation analysis through ConSurf method Tyr77 was observed as highly conserved among others. Molecular docking of endo-1, 4-beta xylanases (1YNA) with substrates viz. xylobiose and beta-D-xylopyranose established Tyr77 residue as conserved in both the binding sites and InterProScan analysis further verified Tyr77 in family 11 glycoside hydrolase motifs with a significantly good score of 1,604. Molecular docking revealed docking scores for Y77 W (−272.50, −171.47) and Y77R (−272.65, −174.71) for xylobiose and beta-D-xylopyranose also signified higher stability against wild type 1YNA (−271.77, −170.78) indicating that the mutant binding residue present in interactions between the substrate and residues. Being a conserved residue Tyr77 was further considered as a target for single-site mutation from which 19 mutants were generated and energy was successfully minimized through GROMOS 43B1 force field. During authentication of results by PROCHECK and ERRAT studies illustrated that no residues are present in disallowed region. Finally, during 3d2GO analysis we accomplished that mutant Y77 W is showing enhanced confidence levels toward hydrolase activity (0.98), molecular function (0.58), proteolysis (0.19), and binding activity (0.17) whereas mutant Y77F demonstrates good improved catalytic activity with confidence level of 0.95.

A study on the specificity versus stability in computational protein design by Bolon et al. (2005), Bonvin (2006) gives various approaches on these issues, whereas few more recent reports published also indicated target protein rediscovery and its functional annotations (Camacho 2005; Zacharias 2003; Jackson 1999; Young 1998) but majority of these reports are on different proteins, their binding sites and these reports did not deciphered family 11 xylanases.

M. V. K. Karthik and P. Shukla, *Computational Strategies Towards Improved Protein*
Function Prophecy of Xylanases from Thermomyces lanuginosus, SpringerBriefs in Systems Biology,
DOI: 10.1007/978-1-4614-4723-8_5, © The Author(s) 2012

However, Al Balaa et al. (2009), Galperin and Koonin (2000) described site-directed mutagenesis in xylanase XYL1p from Scytalidium acidophilum. One noteworthy contribution on mutant of family 11 xylanase from Neocallimastix patriciarum specified the thermostability and substrate selection. However, our studies were conducted on a different species of fungi (i.e. Thermomyces lanuginosus) by means of pioneering computational strategies using some state-of-the-art tools viz. ConSurf, InterProScan, and GROMOS 43B1.

In this work, we report that the interplay of an amino acid mutation and the substrate could lead to enhanced effects on the mutant. This suggests that these effects may need to be reckoned in the engineering processes of protein stability and further exploration of such learning are required to provide novel indication for selection of enzymes.

References

Al Balaa B, Brijs K, Gebruers K, Vandenhaute J, Wouters J, Housen I (2009) Xylanase XYL1p from scytalidium acidophilum: site-directed mutagenesis and acidophilic adaptation. Biores Technol 100:6465–6471

Bolon DN, Grant RA, Baker T, Sauer RT (2005) Specificity versus stability in computational protein design. Proc Natl Acad Sci U S A 102:12724–12729

Bonvin A (2006) Flexible protein-protein docking. Curr opin struct biol 16(2):194–200

Camacho C (2005) Modeling side-chains using molecular dynamics improve recognition of binding region in capri targets. Proteins 60:245–251

Galperin M, Koonin E (2000) Who's your neighbor? New computational approaches for functional genomics. Nat biotechnol 18(6):609–613

Jackson R (1999) Comparison of protein-protein interactions in serine protease-inhibitor and antibody-antigen complexes: implications for the protein docking problem. Protein Sci 8:603–613

Young KH (1998) Yeast two-hybrid: so many interactions, (in) so little time. Biol Reprod 58(3):302–311

Zacharias M (2003) Protein-protein docking with a reduced protein model accounting for side-chain flexibility. Protein Sci 12:1271–1282

Printed by Publishers' Graphics LLC
MO20120724